William Harkness

On the color correction of achromatic telescopes:

A reply to Prof. Chas. S. Hastings

William Harkness

On the color correction of achromatic telescopes:
A reply to Prof. Chas. S. Hastings

ISBN/EAN: 9783337900571

Hergestellt in Europa, USA, Kanada, Australien, Japan

Cover: Foto ©berggeist007 / pixelio.de

Weitere Bücher finden Sie auf **www.hansebooks.com**

[FROM THE AMERICAN JOURNAL OF SCIENCE. VOL. XIX, FEBRUARY, 1880.]

ON THE

COLOR CORRECTION OF ACHROMATIC TELESCOPES:

A REPLY TO PROF. CHAS. S. HASTINGS.

BY WM. HARKNESS.

In the December number of this Journal, pages 48⁄ 485, the distinguished Associate Professor of Physics of the Johns Hopkins University has criticised my theory of the color correction of achromatic telescopes in language which I quote here to avoid the possibility of misrepresenting it; merely adding numbers to the clauses for convenience of reference:

"These results are directly opposed to those of a recent writer in this Journal (Prof. Harkness in the Sept. number, pp. 191–193). But his conclusions arise from erroneous calculations. (I) Not only does his interpretation of his equation (12) imply the manifest absurdity that in a system of infinitely thin lenses in contact its properties are determined by the order of the lenses, but the interpretation is impossible. True A_2 should have an opposite sign to $A_1 + A_3$, but that asserts nothing as to likeness of the latter symbols in sign. (II) Thus n in equation (16) may be negative and consequently his subsequent reasoning is fallacious, for in that case *n* does not have to be infinite to cause equation (27) to vanish. (III) I may add that the origin of the confusion is in making the ratio D ÷ E in equation (9) constant; it may be, and of course should be, indeterminate."

"(IV) Professor Harkness has made another mistake, founded upon inadequate experiment, which has so important a bearing on the theory of the double objective that it should not be allowed to pass uncorrected. His statement (p. 191) concerning the condition for color correction is substantially correct, though in my opinion, it is not self-evident but requires proof. This proof I shall supply in a forthcoming number of the American Journal of Mathematics. (V) His experiment, however (p. 193), directly contravenes this principle, for he finds that the focal plane does not correspond to the minimum focal distance, but to something greater. (VI) The source of error is the introduction of a variable element in the system, namely, the eye, which would adjust itself differently in observing the star and its spectrum. Had the writer used eye-pieces of successively higher power, thus lessening progressively the power of accommodation of the system, with his prism, he would have seen his points y_m and y_n approach until they sensibly coincided; or better still had he formed his spectrum by a grating (such as perforated cardboard) before the objective, instead of by a prism between the ocular and eye, he could not have been misled, since the uncolored image would serve to control the eye."

"(VII) Finally, the fourth conclusion (p. 196) is strictly true, though we are not to conclude, as would seem from the text, that the detriment due to the secondary spectrum depends either solely upon the aperture or varies inversely as the focal length; * * *"

Let us examine this criticism in detail; referring to its clauses, and to the equations of my original paper, by their respective numbers.

Clause I virtually asserts that three quantities can be arranged in two classes otherwise than by putting one in one class and two in the other. To prove this we remark that equation (12) may be written

$$0 = \Lambda_1(b_1 + 2c_1 y_0^2) + \Lambda_2(b_2 + 2c_2 y_0^2) + \Lambda_3(b_3 + 2c_3 y_0^2) \quad (36)$$

For all glasses of which I have any knowledge, *b* is positive, and very much larger than *c*. The latter quantity is sometimes negative; but when this happens, it is exceedingly small. *r* cannot be otherwise than positive. From these conditions it results that the quantities $(b + 2c y_0^2)$ are invariably positive, and therefore the sign of each term in (36) depends solely upon the sign of its Λ. But in order that (36) may be true, one of its terms must have a different sign from the other two; and just because the properties of a system of infinitely thin lenses in contact are independent of the order of the lenses; the choice of this term is arbitrary. Taking advantage of this circumstance to follow the usual practice of opticians, I made the middle lens different from the other two, and wrote

$$- A_2(b_2 + 2c_2\gamma_0^2) = A_1(b_1 + 2c_1\gamma_0^2) + A_3(b_3 + 2c_3\gamma_0^2) \quad (37)$$

But Clause I declares. "True A_2 should have an opposite sign to $A_1 + A_3$, but that asserts nothing as to likeness of the latter symbols in sign."—A statement which is manifestly untrue, unless it can be shown that three quantities can be arranged in two classes otherwise than by putting one in one class and two in the other.

Clause II asserts that n. in equation (16), may be negative. This is absurd, because $n = A_2 \div A_1$, and it has just been shown that the signs of A_1 and A_2 are always similar.

Clause III declares that $D \div E$ should be indeterminate; and that all my alleged errors arise from making it constant. Referring to equations (6), we see that

$$D = A_1b_1 + A_2b_2 + A_3b_3$$
$$E = A_1c_1 + A_2c_2 + A_3c_3 \quad (6)$$

The A's depend only upon the curves of the lenses, while the b's and c's depend only upon the physical properties of the glasses employed. In designing an objective D and E are both so far arbitrary that any glasses, and any curves, may be chosen: but when the objective is completed I certainly do hold that its curves, and the physical properties of the pieces of glass composing it, are constant. If I am right in this, it follows that both D and E, and also their ratio are constant; Clause III to the contrary notwithstanding.

Clause IV admits the accuracy of my statement that an objective is properly corrected for any given purpose when its minimum focal distance corresponds to rays of the wave-length which is most efficient for that purpose: but says the statement requires proof, and is not self-evident. With the law of dispersion assumed in equation (2), the focal curve can have but one tangent parallel to the axis of abscissas; and I did not suppose it necessary to tell the readers of this Journal that the parts of such a curve nearest the tangent line are those adjacent to the point of tangency. That consideration proves my proposition, and it is so elementary that I thought it self-evident. If more than two lenses, and a dispersion formula involving more than two powers of the wave-length, are assumed; I venture to say that the condition for color correction stated above cannot be proved. It may be true in special cases; but in general, the focal curve will have such a form as to give more than one minimum focal distance.

W. Harkness—Color Correction of Achromatic Telescopes.

Clause V involves the assumption that the focal plane must
be tangent to the focal curve at the point where the latter
makes it nearest approach to the objective. No reason is
assigned for this, and I do not believe any exists.

Clause VI virtually asserts that the focal distance of an
objective is a function of the power of its ocular. For all
astronomical instruments carrying filar micrometers, the first
business of the observer is to place the wires accurately in the
focus of the objective. This once done, they are not again dis-
turbed, unless to make some radical change in the instrument.
A dozen eye-pieces may be used in the course of a single
evening; but no matter what their power, when they are
focused upon the wires they are always found to be focused
upon the objective. Hence, the focal plane always coincides
with the wires. But the plane of the wires is fixed; and the
focal curve, as I have defined it, is also fixed. Consequnently,
the points of intersection of the focal plane with the focal curve
are fixed, and the universal experience of astronomers demonstrates that the positions of the points r_m and r_n do not vary
with the power of the ocular.

As *Clause VII* affirms the correctness of my fourth conclusion, it is only necessary to express my thanks for such an
indorsement; but I cannot refrain from adding that, since this
clause rests upon equations condemned by my critic, there may
be people wicked enough to inquire how these erroneous equations finally led to a correct result.

In this connection it is desirable to state that some months
ago I investigated the relations existing in achromatic objectives between aperture, focal length and secondary spectrum.
As the admissible limit of the latter of these elements is arbitrary, it is not possible to fix absolutely the relations between
the other two; but I believe the focal distance should rare
be less than that given by the formula

$$F = (9{\cdot}01a^2 + 1296)^{\frac{1}{2}} - 36 \qquad (38)$$

in which F is the focal distance, in feet; and a the clear aperture in inches. For small apertures, the foci given by this
expression are inconveniently short; while for large apertures,
they considerably exceed those in general use.

Now consider a system of infinitely thin lenses in contact;
and let us inquire how many lenses are needed in the system,
to bring the greatest possible number of light-rays of different
degrees of refrangibility to a common focus, with any given
law of dispersion.

For this purpose we revert to equation (5), which may be
written

$$f'^{-1} = (\mu_1 - 1)\Lambda_1 + (\mu_2 - 1)\Lambda_2 + (\mu_3 - 1)\Lambda_3 + \&c. \qquad (39)$$

the number of terms being unlimited. For the dispersion formula, we write

$$\mu = \varphi(\lambda) \qquad (40)$$

The form of $\varphi(\lambda)$ is regarded as unknown; but there will be no loss of generality if it is developed in a series arranged according to the powers of λ. We therefore have

$$\mu = a + b\lambda^m + c\lambda^n + e\lambda^p + \&c. \qquad (41)$$

in which a, b, c, etc., are constants, and the number of terms may be taken as great as is desired. Also, let us put

$$\begin{aligned}
C &= A_1(a_1 - 1) + A_2(a_2 - 1) + A_3(a_3 - 1) + \&c. \\
D &= A_1 b_1 - A_2 b_2 + A_3 b_3 + \&c. \\
E &= A_1 c_1 + A_2 c_2 + A_3 c_3 + \&c. \\
F &= A_1 e_1 + A_2 e_2 + A_3 e_3 + \&c. \\
&\quad \&c. \qquad \&c. \qquad \&c. \qquad \&c.
\end{aligned} \qquad (42)$$

the number of these equations, and the number of terms in the right hand member of each of them, being the same as the number of terms in the right hand member of (41). Then, by a simple transformation (39) becomes

$$f^{-1} = C + D\lambda^m + E\lambda^n + F\lambda^p + \&c. \qquad (43)$$

This is the equation of the focal curve; λ being the abscissa, and f the ordinate. Its first derivative is

$$\frac{df}{d\lambda} = -f^2(mD\lambda^{m-1} + nE\lambda^{n-1} + pF\lambda^{p-1} + \cdot \&c.) \qquad (44)$$

which, as is well known, expresses for every point of the curve the tangent of the angle made by the tangent line with the axis of abscissas. The number of rays of different degrees of refrangibility, which can be brought to a common focus, will evidently be the same as the number of times the focal plane intersects the focal curve. But the focal plane is necessarily parallel to the axis of abscissas; and therefore the greatest possible number of intersections of the curve with the plane can only exceed by one, the number of tangents which can be drawn parallel to the axis of abscissas. To find these tangents, we equate (44) to zero, and obtain

$$0 = mD + nE\lambda^{n-m} + pF\lambda^{p-m} + \&c. \qquad (45)$$

As λ can never be either zero, imaginary, or negative, we have to consider only the real positive roots of this equation; each of which corresponds to a tangent. To make the number of roots as great as possible, the quantities D, E, F, etc., must be independent of each other: which will be the case when the right hand members of the equations (42) contain as many A's as there are powers of λ in (41) Hence it is evident that the number of real positive roots in (45) will be one less than the number of powers of λ in (41), and we conclude that—

In any system of infinitely thin lenses in contact, the number of lenses required to bring the greatest possible number of light-rays of different degrees of refrangibility to a common focus is the same as the number of different powers of λ involved in the dispersion formula employed.

The method used in deducing this result was adopted because it exhibits clearly the geometrical relations of the problem. The result itself is evident from a mere inspection of equation (43), which cannot possess more real positive roots than it has independent coefficients, D, E, F, etc.

The color correction of an objective depends only upon the form of its focal curve: which form is as much under control as the nature of the case admits when the coefficients D, E, F, etc., of equation (43), are independent of each other. This, taken in connection with what precedes, demonstrates that—

In an objective consisting of a system of infinitely thin lenses in contact, the color correction cannot be improved by increasing the number of lenses beyond the number of different powers of λ involved in the dispersion formula employed.

This result confirms the conclusion of my former paper, in which I used a dispersion formula involving but two powers of the wave-length, and consequently found but two lenses necessary in an achromatic objective. It also throws a curious light upon the general theory of achromaticity. If the law of dispersion had been such as could be expressed by a function involving but a single power of the wave length, there would have been no irrationality of spectra, the mean dispersive powers might have been just what they now are, and yet, Newton would have been right in saying that achromatic telescopes were an impossibility. Conversely, the greater the number of powers of the wave-length involved in the dispersion function, the greater the number of rays of different degrees of refrangibility which can be brought to a common focus : and this, irrespective of any irrationality which may exist in the spectra. With rational spectra, and a law of dispersion involving at least two different powers of the wave-lengths, a pair of lenses would suffice for the construction of a perfectly achromatic objective. In strictness, these statements apply only to objectives consisting of infinitely thin lenses in contact. Possibly they may require modification when the thicknesses and distances apart of the lenses are considered.

The text books teach that the condition of achromatism for two thin lenses in contact is

$$0 = p_2 f_1 + p_1 f_2 \tag{46}$$

in which f_1 and f_2 are the foci, and p_1 and p_2 the dispersive powers, of the lenses. They further teach that it is sufficiently accurate to put

$$\nu = \frac{\delta\mu}{\mu - 1} \tag{47}$$

in which $\delta\mu$ is the difference, and μ the mean, of the refractive indices for the rays D and F. For a law of dispersion involving at least two different powers of the wave-length, these equations will hold; but for a law involving only a single power of the wave-length, they may be satisfied, and yet the system of lenses will not be achromatic. Instead of embodying, these equations are actually independent of, the essential condition of achromatism; which is that at least two rays of widely different wave-length must be brought to a common focus.

I have not had leisure to examine my critic's figures: nor does it seem worth while to do so. My equation (2) represents refractive indices with an accuracy of about four and a half places of decimals, while most of the authorities upon whom he relies only give these quantities to five places of decimals. If this difference of five units in the fifth place of decimals can originate such changes in the focal curve as he supposes, it is evident that trustworthy conclusions can only be reached by using very accurate dispersion formulæ. Cauchy's formula, as written in equation (2), has hitherto been most used; but when compared with the best observations, the residuals, although small, show some constancy of sign. It has recently been claimed* that Briot's formula, which is

$$\mu = a + b\lambda^{\,2} + c\lambda^{-4} + k\lambda^2 \tag{48}$$

represents the best observations, throughout the whole space from the extreme ultra-red to the extreme ultra-violet, within the limits of accidental error. If such is the case, a triple objective may possibly be better than a double one; but my critic's figures certainly do not suffice to prove this. They are founded upon a formula whose independent variable is not the wave-length of the light, but the refractive index of a standard piece of glass; and his Table II, page 432, shows that when compared with observation this formula yields residuals exhibiting as much constancy of sign, and almost the same magnitude, as those given by my equation (2). The use of any independent variable other than the wave-length, is likely to produce erroneous results, and certainly does not tend to elucidate the subject.

Having seen that a dispersion formula involving only three powers of the wave-length suffices to represent the best observations, and remembering that this circumstance limits the number of lenses which can be employed with advantage in an

* By M. Mouton, in the Comptes Rendus, 1879, vol. lxxxviii, p. 1190.

objective to not more than three; we are now in a position to
appreciate the absurdity of my critic's assertion, page 429,
when, enquiring if it is possible to eradicate the secondary
spectrum by increasing the number of lenses in an objective,
he says, "Theoretically, since a new disposable constant for
color change is introduced with each lens in the system, the
answer is evidently affirmative; * * *"

For an objective consisting of more than two lenses, and a
law of dispersion involving more than two powers of the wave-
length, the condition given in my former paper, page 196, for
the best color correction, is no longer applicable. The problem
then becomes very complex, but I am inclined to think that it is
satisfactorily solved by attributing to each element of the focal
curve a mass proportional to its efficiency for the purpose for
which the correction is required, and varying the curve until
its moment of inertia about its intersection with the focal plane
becomes a minimum. It is also probable that this condition
will suffice to determine the relative merits of double and triple
objectives; the focal curve with the smallest moment being the
best.

Finally, it only remains to reiterate that the conclusions of
my former paper are certainly correct to the degree of accuracy
involved in my fundamental equations—that is, for a system of
infinitely thin lenses in contact, and for the law of dispersion
embodied in equation (2). For a different law of dispersion, or
if the thicknesses and distances apart of the lenses are considered, these conclusions may require modification.

Washington. Dec. 29, 1879.